Charles William Nash, Ontario Dept. of Agriculture

The Birds of Ontario in Relation to Agriculture

Charles William Nash, Ontario Dept. of Agriculture

The Birds of Ontario in Relation to Agriculture

ISBN/EAN: 9783742810793

Manufactured in Europe, USA, Canada, Australia, Japa

Cover: Foto ©Klaus-Uwe Gerhardt /pixelio.de

Manufactured and distributed by brebook publishing software
(www.brebook.com)

Charles William Nash, Ontario Dept. of Agriculture

The Birds of Ontario in Relation to Agriculture

THE BIRDS OF ONTARIO

IN

RELATION TO AGRICULTURE

BY

CHARLES W. NASH, TORONTO.

(PUBLISHED BY THE ONTARIO DEPARTMENT OF AGRICULTURE, TORONTO.)

TORONTO:
WARWICK BRO'S & RUTTER, Printers and Bookbinders, 63 and 70 Front St. West
1898.

REPRINTED

FROM THE REPORT OF THE FARMERS' INSTITUTES OF ONTARIO.

1897-8.

THE BIRDS OF ONTARIO

IN

RELATION TO AGRICULTURE.

By Charles W. Nash, Toronto, Ont.

The agriculturist in the Province of Ontario has annually to suffer great loss from the depredations of two classes of enemies, both individually insignificant, but, by reason of their numbers, very formidable. These are insects and small rodents, chief among the latter being rats and all the animals usually classed as mice.

It is very difficult to make anything like a correct estimate of the average damage inflicted upon the farmer by these little animals, but every man engaged in farming knows by sad experience that he continually suffers from their work. The enormous amount of grain they destroy, and the young trees girdled and killed by them are visible to everyone; but the creatures themselves, owing to their nocturnal habits and secretive lives, are comparatively seldom seen. Their enormous increase and consequent capacity for serious mischief is, of course, owing to the fact that man has interfered seriously with the balance of nature, and has thoughtlessly, perhaps, destroyed the principal natural enemies of these creatures. Man himself is almost powerless to stop their ravages to any very great extent. The constant exercise of his ingenuity in trapping, and so forth, results in very little and occupies his time to no purpose. The natural enemies of these animals are gifted with special faculties for their destruction, and so are able to cope with them. Chief among the enemies of this class of farm pests, and the only ones we shall consider now, are the birds of prey. These birds are wonderfully provided by nature with the means to fulfil their part in maintaining the correct balance between the small rodents and the vegetable kingdom. They are in a manner nature's police, and if not destroyed by man would so keep down the numbers of these small four-footed thieves that their plundering would be scarcely noticeable. Our birds of prey may be roughly divided into two classes, the hawks and the owls, the first feeding by day and the other by night. Of the eagles we need say but little. They are now so rarely found in the civilized districts that their influence for good or ill is practically nothing, except upon the game, and of that no doubt, they destroy a large quantity.

HAWKS.

Of the hawks there are eleven species, occurring regularly in this Province in greater or less abundance every season. These are the Marsh Hawk, Sharp-shinned Hawk, Cooper's Hawk, Goshawk, Red-tailed Hawk, Red-shouldered Hawk, Broad-winged Hawk, Rough-legged Hawk, Duck Hawk, Pigeon Hawk and Sparrow Hawk ; there are two or three others, but they are only occasional visitors. Of these eleven, the Sharp-shinned Hawk, Cooper's Hawk, Goshawk, Duck Hawk and Pigeon Hawk are the species which occasionally make raids upon the poultry yards, and which at all times seem to prefer feathered game to either fur or insects ; these should, therefore, be shot whenever the opportunity is given. The Sharp-shinned Hawk and Cooper's Hawk are the two species which most frequently attack the poultry. They are both small hawks, but make up for their lack of size by their boldness and dexterity. It is but seldom that they attack a full-grown fowl, but if they once find an accessible lot of chickens they will continue to visit the flock until they have taken them all, or are killed in the attempt to do so. The mischief done by these two species has been the principal cause of the prejudice existing in the farmer's mind against all the hawk tribe, and is usually given as the excuse for the slaughter of all the valuable species whose constant work inures to man's benefit. The food of the Duck Hawk and the Pigeon Hawk consists chiefly of wild birds, but they rarely visit the farms, their usual resort being the marshes and shores of lakes frequented by water fowl. The Pigeon Hawk is not so named because it has any preference for pigeons, either wild or domestic, but because it slightly resembles a pigeon in shape both when on the wing and when at rest.

The Goshawk fortunately does not visit the cultivated portion of Ontario in any numbers regularly ; it is a winter visitor only, and rather an expensive one to entertain when it does come. The winter of 1896-97 was one of the seasons in which it was particularly abundant throughout southern Ontario, and poultry owners suffered greatly from its destructive powers in consequence. This hawk is a large, powerful bird, quite capable of killing and carrying off a full grown hen. The adult is dark, slaty blue above, blackish on the head, the breast and belly barred pale slate and white with sharp black streaks : the young, dark brown above, white beneath with oblong brown spots ; length, about two feet. Owing to its boldness and strength it is capable of doing a good deal of damage, and should consequently be killed whenever seen. As previously stated, this hawk only occurs in winter, and therefore it is not likely to be mistaken for any of the hawks whose food habits are of benefit to mankind. As a general rule, if a hawk is seen about the farm-yard during the winter it is safe to assume that it is there for no good purpose, and the gun should be brought into requisition at once, as all our beneficial hawks migrate southward when cold weather sets in.

From the above species, all of which are undoubtedly injurious to the interests of the agriculturist by reason of the destruction they work in the poultry yard, and amongst our insectivorous wild birds, we turn to the remaining six species of the hawks frequenting this Province, every one of which spends the greater part of its time and devotes its energies to work the destruction of the animals and insects which are known to be amongst the greatest pests the farmer has to contend with; these are the Marsh Hawk, Red-tailed Hawk, Red-shouldered Hawk, Broad-winged Hawk, Rough-legged Hawk and Sparrow Hawk.

Nearly everyone knows the Marsh Hawk and has seen it gracefully skimming over the low meadows, occasionally hanging poised over one spot for a second or two, and then dropping down into the long grass ; this drop generally means the death of a meadow mouse, sometimes, but more rarely, a frog ; of these two creatures its food principally consists, and the number of meadow mice destroyed by each of these birds in a season must be something enormous. As many as eight have been found in the stomach of one of these hawks, and four or five quite frequently. The hawk's digestion is very rapid, and their hunting and feeding is continued with but few intermissions from daylight until dark.

How many mice each bird would take on the average each day would be difficult to state exactly, but it is safe to assume that at least six would be required. Now multiply that by the vast army of these hawks that resort to this Province and the total number of mice destroyed would be amazing; and then against this good work constantly going on there is no damage to be set off. Not one instance, in thirty years' observation of this bird's habits, has ever come to the writer's knowledge of their having attacked a single domestic fowl. It does sometimes take a meal off a dead duck or other bird it may find lying in the marshes, but it is doubtful if it ever kills for itself a bird of any kind, at any rate in this Province. Every farmer and every sportsman in the land should do his utmost for the protection of this bird. Unfortunately they are constantly destroyed by persons who are ignorant of the good they do, and thousands are killed every autumn by mischievous people who must shoot at everything they see that has life in it. If people who wantonly shoot hawks would sometimes look at the stomach contents of the birds they kill they would soon be convinced of the wrong they were doing and would perhaps exercise sufficient common sense to refrain from continuing the evil practice.

For the sake of brevity the Red-tailed Hawk, Red-shouldered Hawk, and Broad-winged Hawk may be considered together. These three common species are usually known as " Hen Hawks." Why, however, it would be difficult to say. They are all fairly large, slow, heavy flying birds, whose food consists principally of mice, squirrels, toads, frogs and snakes; very rarely do they ever take a bird of any kind. In fact it would be extremely difficult for them to do so, unless the bird was very young, or injured seriously. They will, when pressed by hunger, feed on carrion, but the staple article of diet with them is meadow mice and squirrels, varied, as before stated, by toads, frogs and snakes, besides grasshoppers and other insects.

I have specially omitted from this group, to which it really belongs, the Rough-legged Hawk. This is done purposely, because the great value of the species to the farmer should be particularly pointed out, the bird having been most unjustly persecuted. It is the largest of the Canadian hawks, and one that deserves the greatest consideration and protection from every man having an interest in agriculture. It can be safely said that this so-called " Hen-Hawk " has never killed a head of poultry at any time, nor do they ever kill birds of any sort. During the fall of 1895 these hawks were very abundant in southern Ontario and large numbers were killed. I obtained all the bodies I could for the purpose of investigating the contents of their stomachs, and I spent much time in watching their habits whilst feeding. All day long, every day from the first of October of that year to November 28th, the birds were constantly passing slowly along through southern Ontario, feeding as they went, and not one fowl was taken or attacked by them anywhere, so far as I could learn, and I made enquiries from poultry keepers wherever I could. In all, 32 specimens were examined by me, and the result corroborated my experience during the last thirty years. In one stomach I found a frog, in another the flesh of a muskrat—taken from a pile of bodies of these creatures which had been thrown together in Ashbridge's Marsh. Another stomach was filled with large grasshoppers, and the rest contained mice, and nothing but mice, or traces of them, ranging in quantity from a little fur and a few bones to seven whole ones. From this it can be judged whether or not the Rough-legged Hawk is the friend of the farmer.

The attention of the Department of Agriculture at Washington was some time ago called to the fact that mice and other destructive rodents were largely increasing throughout the United States, and it was suggested that the constant destruction of the hawks and owls was the reason for it. In consequence of this the Department placed the matter in the hands of Dr. Merriam and Dr. Fisher, two of the leading ornithologists of America, with instructions to prepare a report on the subject. This they have done, and the result of their investigations, which I shall give at the end of this chapter, shows conclusively that all the hawks which I have referred to as being beneficial to agriculture are of the greatest possible value in ridding us of enormous numbers of destructive animals, and that they are practically innocent of the commonly urged charge against them of poultry-killing.

There is only one more species of hawk to be considered, and that is the beautiful little Sparrow Hawk, probably the commonest of all our hawks, and which may be distinguished from any of the others by its smaller size and red back. It may be constantly seen hovering over fields in Ontario, all through the summer, for it breeds with us, raising its young in a convenient hole in a tree, frequently choosing one that has been deserted by one of the large woodpeckers. The very small size of this bird precludes the idea that it can take a full grown fowl or even a pigeon, and I have never known in my own experience that it has ever taken a young chicken. Its principal food consists of mice and grasshoppers, of both of which it consumes immense quantities, but it does occasionally take wild birds, more particularly those which frequent the open fields and skulk in the grass or run about the stubbles. The birds taken by these species are, however, so few in number compared to the number of mice which it destroys, and the good it does in reducing the swarms of grasshoppers which infest our fields, that we may well forgive its slight trespasses, the balance of good over evil being so great that the birds deserve our protection. The following shows the result of the investigation made by Dr. Fisher at the request of the Department of Agriculture of the United States :

Red-tailed Hawk. 562 stomachs examined : 54 contained poultry or game birds ; 51, other birds ; 409, mice and other animals ; 37, reptiles, etc. ; 47, insects ; 9, crawfish, etc. ; 13, offal ; and 89 were empty.

Red-shouldered Hawk. 220 stomachs were examined : 3 contained poultry 12, other birds ; 142, mice and other mammals; 59, reptiles, etc. ; 109, insects 7, crawfish ; 2, offal ; 3, fish ; and 14 were empty.

Broad-winged Hawk. 65 stomachs were examined : 2 contained small birds ; 28, mice and other mammals; 24, reptiles, etc. ; 32, insects, etc. ; 4 crawfish ; and 7 were empty.

Rough-legged Hawk. 49 stomachs examined : 45 contained mice and other mammals ; 1, lizards ; 1, insects ; and 4 were empty.

Sparrow Hawk. 320 stomachs examined : 1 contained a quail; 53 other birds; 101, mice and other mammals; 11, reptiles, etc.; 244, insects, etc.; and two were empty.

Marsh Hawk. 124 stomachs examined : 7 contained poultry or game birds; 34, other birds ; 79. mice and other mammals ; 9, reptiles, etc. ; 14, insects ; and 8 were empty.

Thus it can be seen that of the 49 stomachs of the Rough-legged Hawk examined by Dr. Fisher, and the 32 examined by me, in 1895, not one contained a trace of any domestic fowl and nearly everyone contained mice. Yet many people persist in calling this bird the " Big Hen Hawk " and in treating it as an enemy, when both by law and public opinion it should be protected by every possible means. The statement as to all the other species that I have referred to as beneficial is equally corroborated by my own experience, and shows how well entitled these birds are to consideration at our hands instead of the persecution they usually meet.

OWLS.

For some reason owls have always been treated with a certain amount of ridicule and contempt. In the minds of the ignorant and superstitious they were associated with cats and witches, and were supposed to possess a certain amount of influence with the latter, whose orgies they entered into with a good deal of spirit. In mythology, however, this bird was treated respectfully. Minerva, the goddess of wisdom, selected it as her attendant, and "as wise as an owl" has passed into a proverb by reason thereof.

Most of the owls seen in the day-time appear to be stupid, clumsy and inert creatures, as they sit winking and blinking in the unaccustomed light, striving as much as possible to shade their wonderful eyes from the too-powerful rays; but see these birds at dusk and after—what a transformation takes place! They are then as alert as any hawk; their soft plumage enables them to skim noiselessly around our farm buildings and over the fields in search of their food, unlucky then is the mouse or rat that ventures to show itself, or even utter a squeak from its hiding place in the grass, (for an owl's ears are as wonderfully constructed as its eyes, and their hearing is as acute as their sight). The fate of that mouse will be sealed, and it will vex the farmer no more.

Some of the owls however, are day feeders—the Snowy Owl and the Hawk Owl I think entirely so—while the Great Horned Owl seems to be almost as active on dull days as at night; and whether the day be bright or dull these birds can always see well enough to take care of themselves and keep out of the range of a gun. In the cultivated portions of the Province of Ontario we have five species of owls that may be treated here as residents. They are not strictly so, as there is a certain migratory movement amongst them, caused probably by the failure or abundance of their food supply, which may cause them to either leave certain districts for a time or gather there in larger numbers than usual. Many instances are on record of plagues of mice having been stayed and the trouble removed by the arrival on the infested spot of large numbers of owls; these birds rapidly killed off the mice and then scattered again. Our resident species are the Great-Horned Owl, Long-eared Owl, Short-eared Owl, Barred Owl and Screech Owl.

The Great Horned Owl, or "Cat Owl," as it is often called, is the only one I have ever know to attack poultry, and it can work havoc amongst them if they are left out to roost in unprotected places. The destruction of this owl is certainly justifiable and necessary where it has taken up its quarters in a locality in which poultry is kept. It also captures great quantities of our favorite game birds, more particularly Ruffed Grouse, many a brood of which goes to satisfy the hunger of the Horned Owl's family, and are so lost to the sportsman. But as against the charge of poultry and game killing which has been proven against it, this owl has some redeeming qualities. It kills great numbers of rats, mice, squirrels and other rodents that are injurious to farmers, and strange to say it seems to be a determined enemy to the skunk. Numbers of cases have been cited in which the flesh and hair of this animal have been found in the stomachs of these owls, more particularly in the spring, and I know that fully one-half of the bodies of these birds that I have handled, were well perfumed with the odor of skunk—in many cases so much so, that I have had to throw away many fine specimens the smell being quite unbearable. Possibly these birds are fond of strong odor, for those whose feathers are not scented with skunk perfumery, have generally a strong odor of muskrat, the flesh of which they also appreciate. I have frequently known them to hunt and kill these rats in the spring, during the day time when they were about the banks of the creeks, driven there by the high water of our usual spring freshet. These owls are very powerful birds, usually killing for themselves all the food they eat, and only resorting to carrion in the direst extremity of hunger. Turkeys and guinea fowls, from their habit of roosting in trees, frequently fall victims to the strength and rapacity of these creatures. In such cases only the head and neck of the slain will be eaten, the bodies being left to animals of less power, or meaner ambition, to finish.

The Long-eared Owl is a much smaller bird than the last (being about fifteen inches in length), and contents itself with much humbler fare than its big cousin. It is fairly common throughout the cultivated districts, particularly in the autumn, when it may often be found in clumps of willows and alders that have been left in the low places about the fields and pastures. Quite frequently a pair will be found together. These are not, however, always male and female. I have never seen any evidence to show that this owl ever attacks poultry, and I do not believe that it could kill any domestic fowl larger or stronger than a pigeon. Its chief food consists of mice, varied occasionally by small birds and insects, more particularly the wood-boring beetles; of these one or more will generally be found in the stomach or crop of every specimen examined. It is nocturnal in its habits, rarely moving about during the day unless disturbed, and even then it seems loth to move. Only once have I seen it attempting to hunt in daylight, and that occurred in western Ontario on a very dull, still day in November, when about four o'clock in the afternoon I saw a pair of them hovering over a field of long grass into which we had driven a bevy of quail. I suspected the owls of quail-hunting on their own account so followed them and shot both, but their stomachs contained no trace of feathers—nothing but mice. The only harm these owls can ever justly be accused of doing is the occasional killing of a small bird, and that is so far overbalanced by the great amount of good they do, that they are entitled to all the protection possible.

The Short-eared Owl is about the same size as the last named species, but may be distinguished from it by the absence of the long ear-tufts, which are a conspicuous feature in the latter. This is probably the most abundant of all our owls, but it seldom frequents cultivated land, usually resorting to low-lying meadows and marsh hay lands. It is most commonly seen in the autumn, and appears to be somewhat gregarious, large numbers sometimes arriving at one of their feeding grounds together, and remaining there for a few days, then all move off again as they came, to be replaced after a short interval by another lot. The great bulk of them leave this Province by midwinter, or before if the snow should become deep, their movement towards the south being regulated entirely by the depth of snowfall. Whilst the ground is uncovered they are able to obtain a full supply of mice, which form the staple article of their diet; when the snow is deep the mice work underneath it. The supply being cut off, they are driven southward, whither the small birds have already gone, so they cannot fall back upon them. Unfortunately this is a bad failing with the Short-eared owl—in fact my experience shows that it feeds upon mice and small birds indiscriminately, and what is worse, I am satisfied that it kills far more birds than it can eat. Near my home there is a large marsh partially surrounded by low meadows, which support a rank growth of grass, rushes and weeds of various kinds. This place is much frequented in the autumn by sparrows and warblers, migrating southward; in fact at times the place fairly swarms with them. Suddenly a number of Short-eared owls will appear on the scene, and then numbers of small birds will be found lying about dead, some partly eaten and others with only the skull crushed and a few feathers plucked off. At these times I have shot many of the owls, and found the crops and stomach to contain mice and small birds mixed. This will go on for a few days, or until the owls leave, and each morning the number of dead birds lying about will have increased. After the owls have gone the destruction ceases, only to begin again when the next lot of owls arrive. The small birds thus destroyed are of the greatest value to an agricultural community, and their loss is much to be deplored ; but on the other hand the owls destroy an immense number of mice, so that the good they do probably balances the evil, and in such a case the best way is to let nature take its course without our intervention.

The *Barred Owl* is so rare with us that its influence on agriculture, either for good or ill, is practically nothing. The few I have found in this Province have always contained mice, but to the south of us, where the poultry are allowed to roost in trees, it is charged with occasionally killing half-grown chickens.

The noisy little *Screech Owl*, that may in some winters be found in half the barns in the country, is well known to every one, and should be protected by every farmer. It watches the granary, the barnyard and the garden, and is the most indefatigable mouser we have. It seems not only to kill mice for its immediate wants but also for the pleasure of hunting them. If the roosting place of one of these birds is examined after the bird has used it for a short time, numbers of dead mice will be found, most of them untouched after being killed and deposited there : probably they lay up this store in order to provide against nights of scarcity, but in nearly all cases it will be found that they are well ahead of any danger of famine. Not only does this little owl rid the country of numberless mice but in towns and cities it does useful work in keeping the common House Sparrow within proper limits. During the winter particularly, it may often be seen hunting about verandahs, under eaves and among the Virginia creeper growing around dwelling houses, for the sparrows that roost there, and it will go regularly over the same beat night after night, until the accessible sparrows are thinned down, so that it finds it more profitable to change its hunting ground. Besides its great value as a destroyer of mice and House Sparrows, the Screech Owl eats large numbers of large beetles, particularly the wood-borers and May beetles, both of which classes of insects are capable of doing great injury if suffered to become too numerous. Grasshoppers also form a considerable article of this bird's diet. The good qualities of this little owl cannot be overestimated. Its food consists entirely of such creatures as are most injurious to the crops, and it has not a single evil habit. It should, therefore, be carefully protected and encouraged to take up its abode in and about the farm buildings. This I believe it would readily do if it was left unmolested. All it asks in return for its valuable services, is peace and quiet, and a dark corner to roost in during the day.

The Great Gray Owl, the Snowy Owl, the Hawk Owl, Richardson's Owl and the Saw-whet Owl are only irregular visitors, usually occuring in the winter. The two first named are large birds whose food consists chiefly of game birds when in their northern home : here they feed upon the small rodents.

The island and sandbar to the south of Toronto is usually visited by a few Snowy Owls every winter. Here the birds feed upon the common house rats which are altogether too abundant at this spot. As every owl of any kind that visits the place is at once shot the rats, having it all their own way, are increasing rapidly.

The Hawk Owl hunts by day on the prairies of the Northwest, and where it occurs in sufficient numbers it must do much good by the destruction of meadow mice. Its visits to us are so rare, however, that it need not be considered here.

Richardson's Owl and the Saw-whet Owl are two little owls that destroy many mice and noxious insects, but are too rare to need further mention.

Of the ten species of owls before mentioned, nine of them are among the best of the farmer's friends, watching and working when he is sleeping. In following out the natural law which governs their lives they greatly help to keep in check that vast army of little animals which, if allowed to increase unrestrained by their natural enemies, would in a few seasons destroy all vegetation on the face of the earth. The chief and most effective check upon the undue increase of this army of rats, mice, etc., are the birds of prey. These birds are endowed with natural faculties specially adapted for the work they do, and they do it well, the only trouble is that we have too few of them. If, however, public opinion can be brought to bear on this important matter before it is too late, and the wanton and useless destruction of our beneficial hawks and owls stopped at once, the balance of nature may be restored, to the great advantage of mankind.

The following shows the result of Dr. Fisher's investigation of the food habits of the owls as reported to the Department of Agriculture at Washington.

Great Horned Owl. 127 stomachs examined : 31 contained poultry or game birds : 8, other birds ; 13, mice ; 65, other mammals ; 1, insects, etc. : 1, fish, and 17 were empty.

This shows that althought the birds does some injury by its raids upon game and poultry, yet its evil propensities are somewhat counterbalanced by its destruction of mice, rats, rabbits and other small mammals. It is the only one of the owls about whose record for good there can be any doubt. All the others should be protected, while this one should certainly be killed off, if it takes to visiting the barnyard.

Long-eared Owl. 107 stomachs examined : 1 contained a game bird ; 15, other birds: 84 contained mice ; 5, other mammals ; 1, insects, and 15 were empty.

Short eared Owl. 101 stomachs examined : 11 contained small birds ; 77 contained mice ; 7, other mammals; 7, insects, and 14 were empty. My own experience shows a larger proportion of small birds than the above.

Barred Owl. 109 stomachs examined : 5 contained poultry or game birds ; 13, other birds ; 46, mice ; 18, other mammals ; 16, frogs, lizards, etc. ; 16, insects; etc., and 20 were empty.

Screech Owl. 254 stomachs examined : 1 contained the remains of a pigeon 38, other birds ; 91, mice ; 11, other small mammals ; 25, frogs, lizards, etc.; 107 insects, etc., and 43 were empty.

The above examinations of the stomachs of our resident species show most positively that, with the exception of the Great-horned Owl, the whole family are of the greatest value to the farmer. My own experience, both in Manitoba and Ontario, corroborates this, and is perhaps a little more favorable to the owls, for (always excepting the Great-horned Owl) I have never found a trace of a game bird or domestic fowl in any of them.

CROWS, BLACKBIRDS, ORIOLES, ETC.

In this chapter I will deal with two families of birds, both of which are charged with being amongst the worst of the feathered enemies of the farmer. The mischief they do is plainly visible ; the good not always seen. When the crow visits the corn field in the spring, and is seen digging into the hills, abstracting the half-sprouted grain, and when the blackbirds in clouds alight on the ripe wheat and oats, eating much and threshing out more, so that it is lost to its lawful owner, it is not to be wondered at that the farmer loses his temper and says in his wrath that all birds are a nuisance ; but these birds also do some good, though, as they have not acquired the knack of advertising it, their benefits are quite overlooked. If their case is tried impartially it may be found that even the Crow, like another celebrated personage, is not quite " so black as he is painted." I do not think the merits of the crows, or any of the so-called blackbird family, will be found sufficiently great to entitle them to protection, but their faults scarcely warrant their extermination, except in the case of the cow-bird, to be spoken of hereafter.

Crow. Twenty-five years ago the Crows of the Province of Ontario were as regularly migratory as the Robins. A few occasionally stayed through the winter with us, and their doing so was considered a sign that we would have a mild season. As the land has been brought under cultivation, and more particularly in neighborhoods where market-gardening is carried on extensively, the number remaining through the winter has steadily increased, so that the species may now be considered a resident one. In the vicinity of Toronto vast flocks gather at the close of the autumn, feeding on the refuse vegetables left in the market gardens outside the city, and resorting at night to some of the pine woods still left standing. In these they roost all through the winter. They may sometimes be pinched by hunger, but, unless the snow becomes too deep, they can generally get at the piles of manure drawn out on the market gardens, and other refuse left about the land. At this time they do no harm, and probably a little good, as they pick up many mice and insects in their foraging, but when spring opens they again scatter over the country and seek their nesting places. Seeding operations are now going on, and the first of the Crow's mischievous propensities asserts itself as soon as the grain has absorbed sufficient moisture from the ground to become soft and has slightly sprouted ; then it becomes a favorite morsel for the crows. Corn is preferred to any other grain. I have rarely found any quantity of any other grain in the stomach of a Crow, and even when the birds have been seen feeding among the hills of sprouting corn and have been shot right on the spot, I have always found the stomach contained quite as large an amount of insect remains as of corn, the cut-worm forming one of the Crow's choicest articles of diet, and the question arises as to whether it is not better to let the Crow have a little corn and get rid of the cut-worm, than to let the cut-worm take off a lot of corn if we get rid of the crow. Later on I will say something about the history of this same cut-worm. It is always wisest " of two evils to choose the least," and it must be conceded that the corn-eating propensity of the crow is an evil ; but it is certainly less than the evil done by the cut-worm. So perhaps, so far as the Crow's case goes here, it would be as well to call the balance even and give the Crow the benefit of it.

The next scene in the Crow's proceedings shows him with a lively and decidedly hungry family of four or five little ones, whose cravings demand constant attention from their parents ; the variety of food supplied to these insatiable youngsters will vary somewhat according to the locality in which they are placed ; in any case, no more grain will be taken by the parent birds ; their food now will consist entirely of insects, mice and the young of other birds. Nor will they stop at the young if they can catch an adult small bird. Sometimes they will try to elude the viligance of an old hen and will snatch up her chickens more adroitly than any hawk : ducklings fall easy victims to their cunning. It is at this season they do the greatest amount of mischief, by destroying the nests and young of more valuable birds, particularly of such as nest upon the ground. For this reason chiefly Crows should be kept within proper limits as to numbers. Of late years they have increased altogether too fast, and our small birds have suffered in consequence.

After the young birds leave the nest they move about with their parents and feed on a most varied diet. They will make a raid on the fruit grower, and demolish his cherries or raspberries if the idea strikes them, or they will prowl along the lake shore and enjoy themselves for a few days on fish fare, after which they will visit a pasture field and clear out all the wire worms, grubs and mice they may find there : in fact, very few things come amiss to them, as they roam about through the country until the cold nights warn them to get together in some place where they can get at least a bare subsistence to carry them over the winter.

As I have said before, Crows have increased too fast of late years, and we have now too many of them in the country : their numbers can easily be reduced if a little attention be paid to the matter in the spring. Just at nesting time they are less shy and wary than at any other season and can be approached in the trees within shooting distance. If one of each pair was shot off their numbers would soon be reduced to such an extent that the damage they could do would not be noticeable. These birds are so well able to take care of themselves that even more stringent measures might be adopted against them without any danger of extermination, their natural enemies being very few, and those, of that class against which man has carried on a most successful war. Of these the Great-horned Owl was the most noteworthy, but the Great-horned Owl will kill the poultry of a farmer who allows his fowls to roost out on winter nights, and so the Owl must go and the Crow has one enemy the less.

Raven. This species only occurs in the more northerly portions of the Province, having retired before the encroachments of civilization. To the pioneer it is sometimes a nuisance, poultry and young lambs falling easy victims to this bird's strength and rapacity. They also destroy a large quantity of game, but fortunately their number is so small, and the birds themselves so conspicuous, that it is not difficult to get rid of them.

Blue Jay. It is a pity that so beautiful and interesting a bird as this should be possessed of such mischievous propensities as it is, but I am afraid that neither its good looks nor its good acts can be said to balance its evil deeds. This bird, like the common Crow, seems to forget its usual shyness when spring arrives and will leave its wooded haunts and build its nest in gardens, orchards and shrubberies, close to houses and quite within reach of every person passing, nor does it affect any sort of concealment as a rule. I have seen many nests so placed that they were visible from public roads where people and vehicles were continually passing. The female could quite readily be seen sitting, yet the birds carried on their duties regardless of prying eyes. It seems a pity that their confidence should be abused, but I am compelled to say that in all cases that came under my observation the Blue Jays badly repaid the persons in whose gardens they were protected and allowed to raise their young. In the first place they steal a large amount of small fruit, and, further they rob and destroy the nests and young of other birds to such an extent that they are positively injurious to agriculture, the birds they destroy being all of that class whose food consists principally of insects, and without whose assistance I doubt if we could succeed in raising any crop to maturity.

The Blue Jays themselves, however, destroy no inconsiderable number of insects, and they do no damage to grain ; they may occasionally pick off a little corn from the cob, but that is about the extent of the injury they do in that direction. Their unfortunate fondness for the young of other birds more valuable than themselves makes it necessary that they should be destroyed when they take up their residence about our gardens, for it is there, and in our cultivated fields, that our insectivorous birds do the most good, and to get them there we must give them as much protection as possible from their natural enemies, and teach them that they are in greater safety near our dwellings than they would be in the woods. Birds of all kinds soon lose their fear of man if unmolested by him, and particularly if they find that in his immediate neighborhood they can raise their young safely. I know of several farms and large gardens where birds have been encouraged and protected from their enemies; to these places they return in increased numbers year after year, until nearly all available breeding places are taken up. On these premises the owners rarely suffer from the depredations of cut worms or other insects, and so find themselves well repaid for the little care they require to exercise on behalf of their feathered friends.

Bronze Grackle, better known throughout the country as the "crow black-bird," is, when in full plumage, a very handsome bird, and may be distinguished from the other so-called blackbirds by its large size and the brilliant metallic lustre of its feathers. Like the Rook of Europe, it breeds in colonies, and is gregarious at all times of the year. To the farmer, the fruit grower and the lover of birds generally, this bird is a nuisance ; all that can be said in its favour is, that it is very beautiful and that it does, at times, eat a large number of cut worms, for which it may often be seen working industriously on the lawns and grass fields near its nesting place, but as against that it has a heavy record of crimes to answer for. They are early migrants, arriving here about the end of March and resorting at once to their nesting places. From this time until the oats are sown they probably feed entirely on insects, but as soon as the grain is in the ground they visit the newly sown fields and help themselves liberally, varying their diet by taking as many small birds' eggs and young as they can conveniently get at. I have on several occasions seen them attack and carry off young Robins, in spite of the vigorous defence set up by the victim's parents and all the friends they could summon to their assistance. The row made by the despoiled nest owners on these occasions, together with the frantic dashes they made at the robber, would be sufficient to shake the nerves of one of the hawk family, but the Crow blackbird disregards it all, and goes off with its prey.

As soon as the strawberries, cherries, etc., are ripe these birds display a fondness for fruit, and a persistency in gratifying it, that is maddening to the fruit grower, whose profits dwindle day by day by reason of the visits of these thieves, who will continue to carry it off until the young leave the nest. When the young Grackles can fly they gather in large flocks and roam about the country all day, roosting together in vast numbers in some marsh every night. The Dundas marsh, near Hamilton, used to be much favored by them for this purpose ; it is at this season they do the worst of their mischief to the fields of wheat and oats. Not only do they eat an immense quantity, but as they flutter and struggle in their efforts to balance themselves upon the straw of the standing grain, they thresh out and cause the loss of much more. Nor does the cutting and shocking stop their ravages : they still continue to feed upon it, until the last sheaf is in the barn. In the Province of Manitoba where these birds are abundant, I have seen all the grain threshed out from the ears for a space of ten yards in width, round fields which had been selected by them for their feeding ground. In this Province they are rarely to be found in sufficient numbers to do as much damage as that, nor are they likely to become so, for although their chief natural enemies, the hawks and owls, have been too much reduced to be able to keep them entirely in check, yet their number is still manageable and may be kept so by the judicious use of the gun. I advise anyone who shoots them, particularly in the early autumn, to try blackbird pie. Whoever does so will, I think, want to repeat the experiment.

Rusty Grackle. This is a much smaller species than the last and is not of any importance to us from an agricultural point of view. I merely mention it as it occurs here in considerable numbers for a short time in the autumn, but as it does not arrive until the early part of September, the crops are safe from its ravages. In Manitoba, where it is very abundant, it unites with the other blackbirds and destroys a large amount of grain. A few pass through this Province in the spring on their way to the north to breed, but they make no delay and are not noticeable.

Red-winged Blackbird. From an agricultural standpoint this bird has little to recommend it, but to the lover of nature its beautiful coloring and cheery note in early spring render it an object of interest. They are among our earliest migrants, arriving about the middle of March, and resorting at once to the marshes, in which they remain until after the young are able to fly. While in the swamps their food consists almost entirely of aquatic insects, of which the larvæ of the dragon flies form the principal part. As these larvæ form an important item in the food of some of our most valuable fish, and the mature dragon flies feed largely on mosquitos and other small winged insects, the blackbirds are not doing mankind a particularly friendly service by destroying them. This would perhaps not be worth sufficient consideration to warrant our interference with the birds, were it not for their other and more serious failing. As soon as the young are able to fly strongly, which is about the middle of July, they leave the marshes in which they were bred, and in great flocks resort to the grain fields, where, like the grackle, with which they frequently associate, they do much damage, particularly to oats, which they seem to prefer to any other grain. As these birds are very abundant, the loss caused by their plundering must be very great, but they can fortunately easily be managed if a little attention is paid to them in the spring, when they may be shot off on their breeding grounds.

After the grain is carried, they again return to the marshes and gorge themselves on the wild rice, until not a grain of it is left, thereby depriving the wild ducks, etc., of a most attractive food. As soon as the first frost comes they retire to the south, where they cause much worry to the rice-grower. Little can be said in extenuation of these serious faults. They never interfere with other birds or their nests, and they probably destroy some noxious insects, such as cut worms, etc., in meadows, lying near the swamps they frequent in the early part of the season, but this is all that can be urged in their favor.

Cow Bird. Male in summer, all over, except the head, a lustrous, glossy black; the head glossy chestnut. Female and young—dull, sooty black. Length of male, about seven inches, female rather smaller. This bird should be known to everyone, and should be destroyed whenever the opportunity occurs. It is the only feathered creature against which I would advocate a war of extermination, and this I do, because it is not only of no value in itself, but the rearing of each one of its young means a loss to the country of an entire brood of one of our valuable insectivorous birds. It is true that during the early part of the season it frequents the pasture fields where cattle are grazing, and feeds principally on the insects affecting such places, but this is easily counterbalanced by the grain it destroys later on. These birds do not mate, nor do they build a nest for themselves, but the female deposits each of her eggs in the nest of some other small bird The egg is whitish, thickly covered with greyish brown dots. I have found the eggs of this bird in the nests of nearly all the sparrows, finches, and warblers that breed in the Province. After the egg of the Cow bird is deposited, the female takes no further interest in the matter, but leaves it to be hatched by the real owner of the nest in which it has been placed; in due time the young will appear and then the trouble arises. In a few days the young Cow bird has far outgrown its fellow nestlings, in size, strength and voracity, so that it requires and manages to get all the food the parent birds bring to the nest, the result being that the proper occupants of the nest are either starved to death or crowded out by the interloper, who from that time until it is full grown taxes to the utmost all the energies of its foster parents to supply its voracious appetite. Nothing can be more pitiable than the plight of a pair of small birds upon whom one of these parasites has been foisted. They are forced to raise an ugly foundling instead of their own young, and then by reason of the long continued helplessness of their foster child, they are prevented from raising a second brood; for although it quickly grows large and strong enough to crowd out its fellow nestlings and its body develops rapidly, so that it can leave the nest and follow its foster parents through the trees, yet its energy does not develop proportionately with its body, and it requires to be fed for a longer period than the young of any other small bird. The destruction of the natural enemies of this bird, and the constantly enlarging area of cultivated land, both operate favorably for the increase of this pest, so that of late years it has become altogether too abundant. Last year (1897) in the southern part of Ontario it swarmed everywhere, and I noticed an egg of this bird's in quite half the nests of other small species that I chanced to find; of course, in every case I took it out and promptly smashed it, thereby saving the proper brood. It is to the increase of these creatures that I attribute almost wholly the decrease which has become so noticeable in our more useful species. Some idea may be obtained of the terrible destruction worked among the valuable species by Cow birds by just noticing the immense flocks of them that occur here in the autumn, and remembering that for every one of those Cow birds, a brood of some other species has perished. Most of our insectivorous birds produce an average of about four young to a brood, and some of them would raise two broods in a season; the deposit of an egg by the Cow bird in a nest prevents the raising of any young at all of its own by the bird victimized. Just how many eggs each Cow bird lays each season is rather uncertain; in all probability four or five are deposited. If that is so, every female Cow bird that arrives here in the spring and is allowed to follow her own method of reproduction, causes the loss of from fifteen to twenty-five of the young of our most valuable birds. In view of the great increase that has taken place in the numbers of this bird of late years, it is not to be wondered at that our other native species are decreasing, and we should take steps at once to regulate matters. Every person on finding a nest of any of our small birds should look over the eggs contained in it, and if one is found therein differing from the others and corresponding to the description of the egg of the Cow bird which I have already given, that egg should be taken out and destroyed. School teachers throughout the country would do well to impress this upon their pupils.

Shooting the females in early spring is perhaps the most satisfactory way of keeping down the number of this most undesirable bird, and I strongly urge everyone who has access to a gun to use it for this purpose, about his own premises; for, as I have already pointed out, every Cow bird killed at this season means the salvation of much valuable bird life and a corresponding lessening of our insect pests.

Bobolink. One of the most familiar sounds of summer in the country is the merry rollicking song of the Bobolink, to be heard at all times in the fields of scent-laden clover; its bubbling notes, poured out in the exuberance of its spirits, seem to express the feeling of joy that pervades all nature in June. The birds arrive here about the middle of May, the males coming a few days before the females. They resort at once to the hay meadows, and remain there through the nesting season which is concluded by the time the hay is ready to cut. Whilst on the farms their food consists entirely of insects, of which the caterpillars that feed on clover form the greater part. These caterpillars are very abundant, and, where they are not kept in check by the birds, sometimes do serious injury, so that apart from its appearance, and its good qualities as a musician, the Bobolink has a claim upon us which entitles it to our best care and protection. After the hay is cut the males lose their black and white plumage, and become like the females and young in appearance, of a yellowish brown color. They then associate in small flocks and frequent the marshes, feeding on wild rice and the seeds of some rush-like plants until the first frosts come, when they retire to the south for the winter.

In the rice growing States these birds are sometimes accused of doing considerable mischief to the planters' crops, but I am inclined to think that the various species of blackbirds which also resort to these States, are the principal depredators, and by reason of their greater abundance do the most of the damage.

Meadowlark. The Meadowlark is a common though, unfortunately, not now an abundant bird on the farm. Some years ago it could be found wherever the land was cultivated, all through the Province, but owing to its size and slow straight flight, which makes it an easy mark for the gunner, its numbers are decreasing very fast. This is a great pity, for it is an exceedingly valuable bird to the farmer. From the time of its arrival here in March until its departure in November it resorts to the cultivated land and grass meadows, feeding entirely on insects, and never indulging in grain or fruit of any kind. All its work being done amongst the crops upon which man expends his labor and to which he is compelled to look for his subsistence, the benefit conferred is direct and should be appreciated. We cannot make any return for the good it does, but we can at least refrain from destroying its life, and exert ourselves a little to prevent others from doing so. The class of insects upon which this bird feeds during the early part of the season is perhaps the most injurious to vegetable life of all our insect enemies. Its food consists chiefly of those known as cut worms, wire worms, etc., all of which work underground for the most part during the day and only emerge from their hiding places at night. By some highly developed faculty the Meadowlark is enabled to locate these creatures in their hiding places, and being provided with a sharp beak of sufficient length for the purpose, is able to drag them out and devour them. Of all the stomachs I have examined prior to July, the principal contents were wire worms, cut worms, and some few other caterpillars and beetles ; later in the season the food consisted principally of grasshoppers. On two or three occasions I have found a few of these birds wintering with us, in the vicinity of market gardens, and being curious to know if at that season they had been compelled to fall back on a seed or vegetable diet, I shot one out of each lot, and I found the birds were in remarkably good condition. Their stomachs contained, however, nothing but insects, chiefly bugs and beetles, which they had probably obtained from manure heaps and the refuse cabbages left in the gardens. These birds build a domed nest on the ground, in grass fields : their eggs and young are therefore liable to be destroyed by Crows, skunks and other vermin, and those that escape their natural enemies are subject to such continued persecution from gunners who ought to know better, that our beautiful and useful Meadowlark is in danger of extermination, unless some effort is made for its protection.

Baltimore Oriole. The Golden Robin, Fire Bird or Hang-nest, as this bird is sometimes called, is of more importance to the fruit grower than the grain farmer, as it gleans its food entirely among the branches, only visiting the ground for material with which to construct its purse-like nest. Its food consists largely of leaf-eating caterpillars and beetles. It is also particularly fond of the moths which frequent the trees for the purpose of laying their eggs: of these moths it devours large numbers, and in this way it materially assists in keeping down the army of leaf eaters which so frequently strip our trees of their foliage. Very few of our birds will eat a hairy caterpillar, but when they eat a female moth before she has laid her eggs they destroy at one stroke a whole brood of these pernicious creatures, and to this work the Oriole devotes itself with great industry. I have on several occasions obtained a brood of young Orioles and hung them out in a cage near my house for the purpose of discovering the nature of the food brought to them, and found that fully one-half consisted of moths: unfortunately I did not keep a record of the number of these brought in any one day, but it was very large, and the usefulness of this bird in keeping down the swarms of destructive caterpillars, by cutting off the source of supply, was clearly exemplified. When the cherries ripen the Oriole displays a certain partiality for fruit, but the small quantity they take may well be spared them, more particularly as it is only in this direction that they levy any toll for their services. The brilliant coloring of the male, his flute-like note, and the ingenuity displayed in the construction of the nest, all commend these birds to the lover of nature, and we could well spare a few cherries for the sake of having them about our gardens, even if their usefulness was less pronounced than it is. In the south-western portions of our Province the Orchard Oriole occurs. It differs from the Baltimore in being smaller and in color being chestnut and black, instead of the orange and black which marks the present species. Its habits are much the same as those of the familiar Baltimore, but it is too rare to have any economic value.

WOODPECKERS, NUTHATCHES, TITMICE, ETC.

The various species which constitute these families have been grouped together, because of certain similarities in their habits, although structurally they differ widely. They are all tree climbers, and obtain the greatest part of their food from the trunks of trees, some of them by laboriously digging out the grubs which bore into the solid wood, others by prying into every crack and crevice of the bark, where they find insects in various stages of development.

Of the Woodpeckers we have in Ontario nine species, namely, the Pileated Woodpecker (better known as the "Cock of the Woods"), the Arctic Three-toed Woodpecker, the American Three-toed Woodpecker, Hairy Woodpecker, Downy Woodpecker, Yellow-bellied Woodpecker, Golden-winged Woodpecker, Red-headed Woodpecker, and Red-bellied Woodpecker. The first three are true birds of the forest, very seldom showing themselves in the neighborhood of cultivation, so that, although their services are of great value to the country, by reason of the constant war they carry on against the borers, which are so injurious to our timber, we need not consider them in this paper. The Hairy Woodpecker and the Downy Woodpecker are two species that almost exactly resemble each other both in habits and appearance, the only material difference being in their size, the Hairy Woodpecker measuring about nine inches in length, the Downy about six inches. Their food, which consists almost entirely of insects, is

2

obtained either by digging the grubs out of the wood, or picking them out of the crevices of the bark in which they hide during the day. Sometimes during the winter I have found the stomachs of these birds filled with the seeds of the hemlock. These seeds seem to form a favorite food with many of our birds at this season ; the berries of the sumach are also occasionally eaten by the little Downy, perhaps for the sake of the small beetles that are always to be found amongst them. These are the only two vegetable substances that I have ever known either of these species to feed upon.

Both these Woodpeckers are accused of injuring trees by boring holes in them to obtain a flow of sap, which they are said to drink. This is a mistake. The bird having the sap-sucking habit is the Yellow-bellied Woodpecker, an entirely different species, of which I shall speak presently. Nature has most perfectly fitted these birds for their task of ridding the trees of the grubs which bore into them. Their beaks are hard, sharp and chisel-like, so that they are enabled to enlarge the holes inhabited by these insects sufficiently to enable them to insert their long, barbed tongue, with which they extract the larvæ from their hiding places. In the winter these birds frequently visit the orchard, garden and shrubbery, and there they do most valuable work, by destroying the chrysalids of the moths that produce the leaf-eating caterpillars. The toughest cocoon ever spun by a caterpillar is no protection against the sharp beaks of these birds, even the strong case which encloses the chrysalis of the large Cecropia moth is soon torn open when found by a Downy Woodpecker and the contents devoured. Ants and borers in the trees are also greedily eaten by both species ; in fact, nothing in the shape of insect life comes amiss to them, that can be found within their reach. The valuable work done by these birds for the protection of our trees should commend them to every lumberman, fruit grower and nurseryman, and though we cannot do very much to protect them from their natural enemies, we can cease destroying them ourselves and discountenance it in others.

Red-headed Woodpecker. This is the most beautiful bird of the whole Woodpecker family, the strong contrast of the glossy black and the white of its body and the brilliant crimson of the head of the adult birds render them very conspicuous objects of the country ; their value from an economic point of view, however, is debatable. From the time of their arrival here in April until the first strawberry ripens these birds feed on insects entirely, and in pursuit of their food they often adopt the tactics of the fly-catchers by mounting to the top of a telegraph pole or bare limb of a tree, and from thence darting out at any passing insect large enough to attract their attention. If the location selected is a favorable one and food abundant, they will remain at the same spot for some time ; but after the small fruits ripen their tastes change, and they then visit the strawberry patches, both wild and cultivated, and cherries and raspberries are also eaten by them, and carried to their young. When the season for small fruit is over they again resort to their insect eating habit, and so far as I have been able to observe, are not in this Province ever addicted to pilfering grain. I have occasionally seen an odd one make a raid on a vineyard and take a few grapes, and once or twice have seen them pick holes in apples, but the habit does not seem general.

There is no doubt that in the spring they do much good by destroying numbers of mature insects which, if allowed, would deposit eggs to produce vast numbers of injurious caterpillars. It is true also that in districts where small fruit is cultivated for profit they do much harm if they become sufficiently numerous. As the case now stands they are too scarce to do much injury and, except when they are too persistent in their visits to a garden or orchard, they may well be left alone. Although these birds are regular migrants, arriving here about the middle of May and leaving in September, I have once or twice met with them in sheltered woods in south-western Ontario in the winter, where the bright plumage showed to great advantage against the evergreens.

The habits of the Red-bellied Woodpecker are very similar to those of the above species and its economic value about the same, but as it only occurs in the south-western counties of the Province, and then in very small numbers, it need not be further considered.

Golden-winged Woodpecker. Flicker, High Holder, Yellow Hammer, Pigeon Woodpecker, and half a dozen other aliases, testify that this is a well known if not always a popular character. Like the last species, the value of this bird from the fruit-growers' standpoint is debatable, but it is not quite so much given to fruit eating as the Red-head, though when it has seven or eight hungry young ones to feed and it finds a cherry orchard handy, it will help itself to a good many cherries, for which it has a decided predilection. Apart from this unlucky habit the bird has many good qualities. In some of its ways, it much resembles the Meadowlarks; like them it may often be seen stalking about on the ground searching for ants, of which it destroys vast quantities. I have often found their stomachs filled with them, and have rarely examined one without finding it contained some of these insects; it also devours great numbers of grasshoppers, beetles, moths and other ground insects. This bird is really a ground feeder, for, though classed among the Woodpeckers by reason of certain similarities of structure, it does less woodpecking than any other of its class, the beak not being as well fitted for that operation as the beaks of the others. It has also the peculiarity of being able to perch crosswise on a branch, a method rarely adopted by its relations. There is one other evil trait I have seen this bird exhibit, on two occasions only, that is the destruction by it of nests of the Bluebird; both the nests destroyed were built by the Blue-birds in holes in trees much higher than usual, probably from forty to fifty feet from the ground. I am not certain what the nests contained at the time, but I saw the woodpeckers pull out the nests and throw them piecemeal to the ground in spite of the resistance of the Bluebirds, but I found no trace of eggs or young; if there were any they must have been eaten. It is probable that the woodpeckers wanted the nesting site for themselves, and so dispossessed the owners. If so they were disappointed, for I settled the question by killing them, but am sorry to say I omitted to examine the stomachs to see whether or not they had devoured the young Bluebirds, if there were any. I am inclined to think these were exceptional cases; they occurred over twenty years ago and I have never seen a repetition of the trick. If these birds become a nuisance in a garden or orchard, they can easily be killed off while they are comitting their offence, but I think that through the country generally, the good they do far overbalances the little damage they may do locally.

Yellow-bellied Woodpecker or Sapsucker. Adult male, crown and chin crimson, back and wing coverts black and white, wings black with a large white bar, tail black, inner web of the two central feathers white with black spots, breast black edged with yellowish, the rest of the under parts dull yellowish, the sides white with black streaks. In the female the crimson of the crown and chin is wanting, the crown is black with sometimes a few traces of crimson on the forehead, the chin is white. I give a description of this species in order that it may be distinguished from the other small Woodpeckers, because it is principally owing to the propensity for drinking sap which the bird has, that a certain prejudice exists in some localities against all the Woodpeckers or Sap suckers as they are called. It is quite true that these Woodpeckers do, in the spring, when the sap is rising, bore small holes in the bark of various trees for the purpose of obtaining the sap as it flows from them, and also to attract the insects upon which they feed to the same spot, so that they can satisfy their hunger and thirst without having to over-exert themselves in so doing. If life was not so short I might be tempted here to go into the question as to whether this bird had to acquire this habit because its tongue was peculiarly fitted for it, or whether the tongue became modified so as to just suit the habit after the bird had acquired it ; for the bird's tongue certainly differs from that of other Canadian Woodpeckers and is admirably fitted for the use to which it is put. A discussion of the question would exceed the scope of this article, and probably not lead to anything after all. We know the bird has this habit and the question is, what is the effect of it upon the trees which are bored ? I have made what observations I could, and as many enquiries from others as possible, and have come to the conclusion that the only real damage done is that a young tree may be tendered unsightly for a time, or it may even be permanently disfigured by some peculiarity in the healing of the bark, but usually no harm ensues. That a tree ever was or could be killed by it I do not believe, for I have never yet seen or heard any evidence in proof of it.

Apart from its sap drinking peculiarity the bird's record is excellent ; it is not a fruit or grain eater, but devotes itself to the destruction of insects that live on the trees or hide in the loose bark. Ants form a large proportion of its food. These it obtains from the rotten wood in which they burrow, as it does not descend to the ground in search of them. Beetles and moths are also sought out and devoured, but as this bird's tongue is not as well barbed as that of some of the other Woodpeckers, fewer grubs of the wood-boring class are eaten by it. I suppose if any man believes that these birds are doing an injury to his trees he should be allowed to protect himself in the only way possible, viz., by getting rid of the birds on his own premises; but for his own sake he should be sure he gets rid of the right one, and that neither the Downy nor the Hairy is destroyed by mistake. Both the Downy and the Hairy Woodpecker remain with us all through the year, whilst the Sapsucker is a summer resident only; so that whenever a Woodpecker is seen in the winter it should be spared, for it is most certainly a beneficial one.

Nuthatches, Chickadee and Tree Creeper. Of these we have two species of Nuthatches,—the White-breasted and Red-breasted,—one Chickadee and one Creeper. They are all resident species, though more frequently seen about cultivated lands in the winter than in any other season. They are among the most active insect destroyers we have, gleaning their food from the bark, branches and leaves of trees, and seldom descending to the ground, though when woodchopping is going on in the bush the logs, sticks and chips will all be carefully searched for grubs which have been exposed by the axe. The familiarity displayed by these little creatures at this time is very pleasing. As soon as work begins and the first few strokes of the axe sound through the bush, they gather round and investigate every piece of bark and decayed wood thrown open, and from each one gather some prizes. It is very amusing to watch the little Chickadee when he finds a large grub of one of the borers partly exposed. He pulls and tugs at it until it comes out, and then securely holding it down with his feet he tears it in pieces and devours it. Without the assistance of the chopper it is but seldom that they can get at the larger grubs that bore deeply into the solid wood, as they have neither the strength nor proper tools for digging them out ; but they have found out that when the farmer gets out his cordwood their opportunity for a feast arrives, and so they take advantage of it. As a general rule, however, they scour the bush, orchard and shrubbery in merry little parties searching for food, from time to time uttering their musical notes, which always have a peculiar "woodsy" quality about them. The seeds of the hemlock are occasionally eaten by the Chickadee and Red-breasted Nuthatch, and the White-breasted Nuthatch is said to sometimes eat beechnuts and acorns, but I have never found any trace of them. The Tree Creeper eats no vegetable substance whatever.

This little group of birds is of the greatest value to the fruit-grower, as they feed principally on the minute insects and their eggs, which are individually so small that they escape our observation until, having seen the damage done by them, our attention is called to their existence, and then it is too late to enable us to remedy the matter for the season.

THRUSHES.

We have in Ontario seven species belonging to this family, all of them migratory, arriving here from the south in early spring and leaving us in the autumn, as cold weather sets in. They are the Wood Thrush, Wilson's Thrush, Grey cheeked Thrush, Olive-backed Thrush, Hermit Thrush, Robin and Bluebird. The Olive-backed Thrush, Hermit Thrush and Grey-cheeked Thrush pass on and raise their young to the north of us; the others remain throughout the summer and breed here.

The Wood Thrush and Wilson's Thrush, or Veery, as it is sometimes called, are strictly birds of the woodlands, and seldom venture far from the edge of the bush, though both species will at times select a garden where there are shrubs for their summer residence, if they find themselves unmolested, particularly if there are no domestic cats about the premises. The cats at all times prefer young birds to mice or rats, and are as much to blame for the decrease of our native birds as bird-nesting boys or anything else, perhaps, except the Cow bird. Wilson's Thrush is one of our most abundant species, but it has the faculty of concealing itself to such perfection that it is often overlooked though there may be many within a few yards of where a person is standing. The Wood Thrush is very rare with us, which is to be regretted, as it is a beautiful songster.

All these thrushes are very valuable birds to the agriculturist, their food consisting for the most part of grubs that live under the surface of the ground, and caterpillars. In the autumn they eat many wild berries, those of the Elder and Viburnum being especial favorites, but they never help themselves to the produce of the farm or garden. The best known and most familiar of the thrush family is the Robin, and opinion is very strongly divided as to its utility. Many fruit growers condemn this bird with great emphasis, stating that it is the worst enemy they have; others weigh its merits and demerits more carefully, and are inclined to think that it at least pays for the fruit it eats by the destruction of insects. No doubt it does take a large number of cherries, strawberries and raspberies, and some grapes, but it is open to question if it were not for the birds whether there would be any cherries, strawberries or grapes, or, indeed, whether any crop could be brought to maturity. The great merit of the Robin is that in the early part of the season it feeds itself and its young almost entirely on cut worms and on the large white grub, the larva of the May beetle. Of all our insect enemies the underground cut worm is about the most destructive, for in feeding it just comes above the surface and cuts off the entire plant, or if the plants are very young and the stems small it cuts off half a dozen or more at one time, only eating a small section out of the stem of each and leaving the plants dead on the surface of the ground. Whole rows of peas, corn, beets, cabbage and cauliflower are often so treated; tomatoes, too, fare badly with them. The only remedy that seems effectual against their attacks is to wrap paper around the stems of the plants from the surface of the soil to the height of about three inches above it. This is obviously impossible in the case of field crops, and it is equally impossible to go over the fields and take the worms out by hand, so that we must rely, for the most part, upon the ground feeders amongst the birds; these are fitted by nature for digging out the insects and devouring them.

Robin. Among the most conspicuous of these birds is the Robin and one need only watch one of them at work in the garden, from April to about the middle of June (which is the season of the cut worm's activity) to be satisfied as to the Robin's good work. I will give the result of an experiment carried on by myself, which will satisfy anyone as to the number of these insects a pair of Robins will destroy when they are feeding a brood of young. In May, 1889, I noticed a pair of Robins digging out cut worms in my garden, which was infested with them, and saw they were carrying them to their nest in a tree close by. On the 21st of that month I found one of the young on the ground, it having fallen out of the nest, and in order to see how much insect food it required daily I took it to my house and raised it by hand. Up to the 6th of June it had eaten from fifty to seventy cut worms and earth worms every day. On the 9th of June I weighed the bird; its weight was exactly three ounces, and then I tried how much it would eat, it being now quite able to feed itself. With the assistance of my children I gathered a large number of cut worms and gave them to the Robin after weighing them. In the course of that day it ate just five and one-half ounces of cut worms. These grubs averaged thirty to the ounce, so the young Robin ate one hundred and sixty-five cut worms in one day. Had it been at liberty it would probably have eaten some insects of other species and fewer cut worms, but this shows near about what each young Robin requires for its maintainance when growing; the adult birds require much less, of course. The average number of young raised by a Robin is four, and there are usually two broods in the season. A very simple calculation will give a good idea of the number of insects destroyed while the young are in the nest. After the young have flown they are apt to visit the small fruit, and it is no doubt very provoking to find a flock of them helping themselves to strawberries, etc. If possible, they should be kept off without destroying them, a resort to the gun being avoided as long as possible.

Bluebird. Twenty years ago the Bluebird was one of the most abundant of the summer residents in the cultivated districts of the Province; there was scarcely a farm throughout southern Ontario upon which two or more pairs of these birds did not breed. The same birds seemed to return regularly to occupy their holes in the old apple trees and fence posts, year after year, and so familiar were they that they actually seemed to know the members of the family whose premises they occupied. In one case, near Niagara, a pair of Bluebirds for several year in succession built their nests in a letter box which was placed at the gate of the farm, opening on the main road. The mail carrier deposited letters and newspapers in the box every day, which were duly taken out by the members of the family. To all this the birds paid no attention whatever, but would confidently sit upon their eggs or visit their young while the box was opened and people stood close to them; and I have seen many similar instances of confidence on the part of these birds.

Of late years the Bluebirds have not remained with us, and they have been much missed. Enquiries are constantly being made as to where the Bluebirds have gone ? That is not so easy to answer, but that they still exist in undiminished numbers I am able to state positively, for so late as last March (1898) I saw many thousands passing over Toronto from west to east. The flight lasted from daylight to nine or ten o'clock every fine morning for about a week. I have seen this same movement every spring for years. My opinion is that the birds have gone back to the new settlements, where they can still find snake fences and pastures in which the old stumps are standing—our modern barbed wire which which has taken the place of the old stake and rider fence having deprived them of a favorite nesting place. The up-to-date fruit grower, too, no longer allows his apple trees to go untrimmed and full of holes, but cuts out the old trees and replaces them with young ones. This has removed many of the old nesting sites, and the birds have spread out over the large area of new country now being brought under cultivation. They introduced themselves to the Province of Manitoba about 1884, and have since become quite common there, having evidently followed the settlers, as they were quite unknown in that country before it was brought under general cultivation. The utility of this bird as an insect destroyer is beyond question. It eats neither grain nor fruit ; occasionally in stormy weather, in early spring when insect food is hard to obtain, it will eat the berries of the sumach, but that is the only vegetable substance I have ever known it to take. The beauty of its plumage, its sprightly spring song and even the rather melancholy farewell notes in which it bade us good-bye, as it drifted southward in the last days of October, made it a great favorite everywhere, and every lover of nature would be glad to see it return and take its old place about the farm once more.

Cat bird. Neither this nor the succeeding species belong to the Thrush family, but there is a sufficient similarity in their food habits to warrant our considering them here. They are closely allied to the famous Mocking Bird of the south, and their musical powers are not very much inferior to that splendid songster. They do not, however, so frequently exercise their power of mimicry. The peculiar mewing note uttered by the Cat bird has caused a certain amount of prejudice to exist against it, and has made it subject to persecution at the hands of most boys ; but apart from this unpleasant note, the Cat bird is one of the most accomplished musicians we have, and it is more to be admired because it does not retire into solitude to pour out its joyous songs, but rather seeks the society of mankind, and in the morning and evening will sing its clear notes from the top of some tree in close proximity to the dwelling house. Its food in the early part of the season consists almost entirely of caterpillars and beetles, which it obtains generally from the branches and leaves of trees, though sometimes after rain it seeks for cut worms and other grubs from the ground. Later in the year it feeds largely upon elderberries and other small wild fruits, and does occassionally levy some slight toll from the garden ; but for all the cultivated fruit it takes it has amply repaid the gardener by its efforts in the destruction of the insect tribe.

Brown Thrush or Thrasher. All that I have said of the Cat Bird applies to this species, but it is not quite so familiar and confiding in its habits. It displays a decided preference for thick shrubbery at some little distance from the house. Here it remains in seclusion for the greater part of the day, but in the early morning and evening the male bird mounts to the top of some tall tree near its haunt, and for an hour or so will sing his beautiful song, which is much louder, though less varied, than that of the Cat bird.

Wrens. This is a most interesting and useful family of very small birds. Four species of them are found in this Province in the summer. Two of them, the

Long-billed Marsh Wren, and the Short-billed Marsh Wren, as their name implies, frequent our marshes and low swampy meadows, where they assist in keeping down the hordes of mosquitos that are bred in such places. The Winter Wren is a more transitory visitor, the great bulk of them only passing through here in the spring and fall migrations. A few, however, remain here through the summer and nest in some secluded ravine in the woods.

The pert little House Wren takes up its abode right in and around the farm buildings, and even in our cities it will find a resting place, if it can get access to sufficient garden room to give it a hunting ground, and as it is quite satisfied to place its nest in a crevice or hole at no great height from the ground, it is not so likely to be dispossessed of its home by the European House sparrow as are birds that prefer a higher location. They are most indefatigable insect hunters, and should be encouraged to build in every garden. All that is necessary is to furnish them with a small box having a hole about one and one-half inches in diameter. Nail this up to a fence or building, about eight or ten feet from the ground, so that cats cannot get at it ; and if any wrens come that way in the spring they are almost sure to take possession of it, and having once occupied it, they will in all probability return every year. The domestic cat is their worst enemy, and they seem to know it, for as soon as they catch sight of one of these detested creatures they start such a scolding that they arouse the whole feathered tribe in their neighborhood. In the autumn they eat a few elderberries, but this is the only vegetable food I have known them to take.

Cuckoos. These birds do not seem to be very well known in our Province, though we have two species, one of which is not uncommon. They are known as the Black-billed Cuckoo and Yellow-billed Cuckoo. Both of them are slim birds about twelve inches in length, of an olive brown color above, and white beneath. The Yellow-billed may be distinguished from its relative by the light chestnut color of the inner webs of part of the wing feathers. This is quite noticeable when the bird is flying. It also has the under mandible of the beak clear yellow. In the Black-billed species, the beak is all black, sometimes showing slight dull yellow marks below. Although the birds themselves are not known, most residents of the country must have noticed the loud harsh notes of ' kow kow " uttered by them, most frequently heard before and during rain; by reason of which the birds are in some localities called "rain crows."

The well known Cuckoo of Europe has the bad habit of laying its eggs in the nests of other birds, but although I have heard our birds charged with the same thing, I have never yet come across an instance of it, but have always found their nesting habits to be quite orthodox, though the nest they build can hardly be considered a model of bird architecture.

These two species of birds are the only ones, that to my knowledge habitually eat hairy catepillars, and of these noxious insects they must destroy a large quantity, an examination of their stomachs generally showing a considerable number of them. On one occasion I found the stomach of a Black-billed Cuckoo packed with the spiny caterpillar of *Vanessa antiopa,* an insect that feeds in colonies and does much damage to the elm and willow trees. The habits of the two Cuckoos are much alike ; the only difference I have noticed is that the Yellow-billed species seems to prefer the upper branches of tall trees in which to obtain its food, while the Black-billed resorts more to orchard trees and shrubbery. I have not found any evidence of habitual fruit-eating against either of them, so that from an economic standpoint they must be considered as purely beneficial, even if they do occasionally deposit an egg in the nest of another bird.

Warblers. This family contains a large number of species, among them being some of our brightest colored and most interesting birds, though none of them are remarkable as songsters. They are all entirely insectivorous, and consequently of great value from an economic point of view. Thirty-one species are known to occur in this Province; of these five are so rare as to be considered accidental visitors. They are the Prothonotary, the Golden-Winged and Hooded Warblers, the Louisiana Water Thrush, and the Yellow-breasted Chat. Probably when they do occur, they remain and breed here. The Cape May, Orange-crowned, Tennessee, Cerulean, and Connecticut are regular but uncommon visitors. Of these the Cerulean is known to breed in some localities in southern Ontario, but it is not generally distributed.

The Parula, Black-throated, Blue, Myrtle, Magnolia, Blackburnian, Baybreasted, Black poll, Palm and Wilson's Warblers all pass on to the north before nesting. Just how far they go is difficult to say, but in all probability the majority of them at any rate will be found breeding in the unsettled districts of Muskoka, Algoma, etc., and some even south of that.

The Black and white, Nashville, Yellow, Chestnut-sided, Pine, Redstart, Black-throated green, Oven bird, Water Thrush, Mourning, Maryland, and Canadian Warblers, are generally distributed and breed with us in suitable localities and in varying numbers each season, the most familiar of them all being the Yellow Warbler which habitually raises its young in and about our orchards and shrubberies. All through the summer they are actively engaged in exterminating the hosts of our smaller insect enemies, and many thousands of broods of caterpillars are destroyed by them before they have become large enough to do mischief.

Flycatchers. These birds, as their name implies, subsist largely upon winged insects, which they capture by darting upon them from some elevated post overlooking an open space frequented by their prey. We have eight species, of which the Crested Flycatcher, the King bird, the Phœbe bird, the Wood Peewee and the Least Flycatcher are summer residents, and the Olive-sided, Yellow-bellied, and Traill's Flycatcher are transient visitors, passing through in their spring and fall migrations.

The King bird is probably the most obtrusive creature of the whole feathered tribe in Canada. As soon as a pair take possession of a tree in an orchard they immediately proclaim the fact to the neighborhood, and then trouble befalls everything wearing feathers that ventures to trespass on what they are pleased to consider their domain. Crows, Hawks, Jays, and Blackbirds are their especial detestation, and should one of these birds appear near their tree an attack by the King birds immediately follows, the assault being kept up until the intruder is ignominiously driven off, having lost a few feathers in the encounter, the loss serving to remind him that other people have rights which he is bound to respect. The King bird captures a vast number of mature insects, both in the air and on the ground, and as at least half these insects would produce eggs to become caterpillars the service rendered is very great. In the early spring, when driven by hunger, the King bird will eat the berries of the sumach, but as the clusters of these berries form a favorite hibernating place for many beetles, it is quite possible that the insects form the attraction and not the fruit. This is the only vegetable substance I have ever known the bird to touch. I have heard complaints from bee-keepers that these birds will destroy bees. It is just possible that they will occasionally take them, but I have seen no evidence that they have acquired the habit. In case the King birds should be seen frequenting the vicinity of hives, it would be well to watch closely before shooting the birds, as they are too valuable to be wantonly destroyed, and in all cases an examination of the stomach contents should be made, and the information gained should be reported.

There is scarcely a farm in the country that has not one or more pairs of Phœbe birds nesting in or about the buildings, and I fancy there are not many bridges of any size under which a nest may not be found ; and so I hope it may continue, for the Phœbe is a most useful and friendly little bird. It has all the good traits of the family without being too aggressive, and no suspicion of any act which is in the least injurious attaches to it. If the birds and their nests are left unmolested, they will return year after year to their old home, and as none of our feathered friends are more valuable than they, we should give them every encouragement to do so.

I have particularly mentioned the King bird and Phœbe because they may be regarded as typical of the whole family to which they belong, and being familiar in their habits, they are likely to be well known to everyone. All the other species are more or less birds of the woods and orchards, but each one of them in its own chosen locality is rendering us good service the whole summer through.

SPARROWS, FINCHES, Etc.

This is a very large family represented in Ontario by thirty-four species. Want of space prohibits my calling attention to the food habits of each of these species in detail. It will, however, be sufficient for the purpose of this paper to refer particularly only to those which in some manner are specially beneficial or injurious to the crops usually cultivated for profit. All these birds are insect eaters in the summer months, and their young while in the nest are fed entirely on insects ; but in the autumn, winter and early spring, the mature birds subsist principally on the seeds of wild plants and forest trees.

The Rose-breasted Grosbeak is one of the largest and most beautiful of the family, and is of more than usual interest because it is one of the very few birds that will eat the Colorado potato beetle and its larvæ, and also the larvæ of the Tussock moth, this last being a hairy caterpillar, very destructive to almost all shade and orchard trees. A specimen of this bird in my possession eats both these insects readily. Unfortunately, these Grosbeaks are of a retiring disposition, and usually resort to the seclusion of the woods, but it is one of the few species that seems to be increasing in Ontario, and if unmolested it may possible become more familiar in its habits. If so, its services in lessening the number of Tussock moths would be of great value. None of the native members of this family are addicted to eating the ordinary grain or fruit crops, but the Purple Finch (the adult male of this species is crimson, not purple) in the spring is sometimes injurious in orchards and gardens, where it destroys the buds of fruit trees. They will also devour great quantities of sunflowers and other seeds. They are not, however, generally sufficiently numerous to cause much loss.

A member of this family about which there has been much controversy, is the imported European House Sparrow. This bird was introduced into Ontario about the year 1873 by some gentlemen who no doubt were under the impression that the sparrows would devote themselves exclusively to killing and eating the caterpillars that infest the shade trees in our towns. They either forgot or did not know that the sparrow belongs to a class of birds whose diet consists of vegetable substance and insects in about equal proportion, and that the Sparrow, having attached itself to the haunts of man, usually obtains its vegetable food from the plants and seeds cultivated by men for their own use. I have read many reports of so-called observers, who have stated that the House Sparrow never eats insects of any kind, that it drives away our native birds, and that it is altogether an unmitigated nuisance. Sweeping assertions of this kind are only conclusive evidence that the so-called observer cannot observe. No one with ordinary perceptive faculties can walk through our public parks, or along one of our streets where there are trees and grass in the summer time, without seeing some Sparrows industriously hunting for insects with which to feed their young, and should anyone have a sparrow's nest under his verandah or about his house in such a position that some of the food brought by the parent birds to their young will fall where it can be seen, the proof that they do eat insects, and in large quantities too, will be very clear. The old birds also eat insects at this season, varying their diet with such undigested grain as they may find in horse droppings, and with bread crumbs and such like refuse from houses.

Sparrows, like the majority of birds, will not eat hairy caterpillars, but I have seen them eat the spiny larvæ of *Vanessa antiopa*, which is one of our shade tree pests that few birds will touch. Besides this I have seen them take moths of almost any kind, including the large Cecropia and Luna moths and the Tussock moth (both the winged male and the wingless female), beetles of many kinds, even such large species as the aquatic *Dytiscus*, which they find on the sidewalks beneath the electric lights to which the beetles are attracted at night, the green cabbage worm (the larvae of the cabbage butterfly)—of these they eat great numbers. They also hunt about fences, and take the pupa of this same butterfly. The currant worms and the mature insects are also taken in large numbers, as are also grasshoppers, and both the black and green aphides that occur on apple trees and rose bushes are eaten greedily. On one occasion a flock of Sparrows completely cleaned off the green aphis from some rose bushes near my windows. It took them several days to finish their work, but they did it effectually in the end.

About harvest time the Sparrows show their grain-eaten proclivities. They then gather into large flocks, and, leaving the town where they were bred, visit the surrounding country and make serious raids upon the wheat and oats, and do more damage while the grain is standing by beating it out than by eating it. It is in early spring, however, that the worst trait in the sparrow's character becomes apparent. Vegetation awakens after the long winter's sleep : the trees put forth their buds, and seedlings break through the soil. The Sparrow, probably needing an alterative after the hard fare of the winter, attacks all these ; nothing green comes amiss to him, and then the gardener, wrathful at the loss of prospective fruit, vegetables and flowers, forgets the good qualities the bird has, and would have the whole tribe exterminated. Whether or not he would be the gainer by this is somewhat difficult to decide. My own opinion at present is, that the number we now have do as much good as they do harm, but that they should not be allowed to increase to any great extent.

The Sparrow is also charged with driving away our native birds. This charge is well founded only in the case of such birds as were formerly in the habit of building in holes and crevices about our houses, such as the Swallows and the Wrens. In the case of the Wrens the difficulty can easily be got over by placing their nest boxes low down, say about eight feet from the ground ; the Sparrows will not then occupy them. But the Swallow problem is not so easy to solve. The trouble arises from the fact that the Sparrows remain here all through the winter and use the Swallows' nests in that season as roosting places. As spring comes they build in them and so have possession when the Swallows return from the south. As they then, naturally enough, decline to turn out, the Swallows have to seek elsewhere for a home ; the result being that we lose a valuable, purely insectivorous bird and get in the place of it one whose value is questionable. Continually shooting off the Sparrows as they appear seems to be the only remedy, and I think eternal vigilance would be required to make it suc · cessful in any place where the Sparrows are well established.

That Sparrows are rather quarrelsome amongst themselves in the season of love-making is evident to everyone, but so far I have not seen them interfere with any other species whose nesting interests do not conflict with theirs. In my own neighborhood, House Wrens, Orioles, Vireos, Cat birds, Wilson's Thrushes, Robins, Chipping Sparrows, Song Sparrows, the American Goldfinch, and the Yellow Warblers have all bred in close proximity to many pairs of Sparrows and have not been interfered with by them ; but if I had not kept a pretty close watch over the nests, and taken out the eggs of the Cow birds which were deposited therein, but few broods would have been successfully raised.

The Sparrow has one particularly good trait which should not be overlooked ; that is, its fondness for the seed of the knot grass or knot weed. This pernicious plant frequently appears on our boulevards and lawns and destroys the grass completely. The Sparrows soon find it out, and small parties constantly visit it and feed upon it, so that it is kept down and in some cases is quite cleared out.

SWALLOWS.

Of this family we have five species, viz : the Purple Martin, the Barn Swallow, Cliff Swallow, White-breasted Swallow and Sand Martin, all regular summer residents. Another one, the Rough-winged Swallow, occasionally occurs here, but as it closely resembles the Sand Martin its appearance is not readily noticed.

The economic importance of these birds is very great; without them the smaller winged insects would multiply to such an extent as to become an unbearable nuisance to men and animals ; for it is, I believe, to these birds chiefly that we are indebted for our freedom in the cleared and cultivated parts of the country from the swarms of midges, black flies and gnats of various kinds that so abound in the woods.

These birds seem to have a great predilection for the society of men, partly because the clearing he makes in a forest country opens up to them the necessary space for feeding grounds, and partly because the buildings he erects affords them convenient nesting places, of which the House Sparrow unfortunately is dispossessing them.

Except in . very stormy weather the Swallows usually capture their food whilst they are on the wing, but in the cold windy days that frequently occur in early spring the insects on which they depend are too chilled to fly, and then the Swallows seek them in open places on the ground. The sandy shores of our lakes are particularly resorted to at such times.

In the latter part of July and the beginning of August the large female ants swarm from their nests, each one prepared to found a colony for herself were she permitted ; the Swallows, fortunately for us, however, interfere and gorge themselves upon these creatures, the Purple Martins particularly, destroying vast numbers of them, even after the ants have divested themselves of their wings; when this has taken place the Martins alight on the ground pursuing them there with the greatest activity.

The Chimney Swift, which closely resembles the Swallows in its habits, except that it never alights on the ground even to obtain the materials for its curiously constructed nest, may be mentioned in connection with them, its economic value being equally great.

Night Hawks. All the Swallow tribe gather their food during the day, and the hotter and brighter it is the more active they seem to be ; the Chimney Swift's period of greatest activity is the early morning and late evening. The Night hawk and Whip-poor-will commence their work at dusk and keep it up till sunrise. Their food consists, for the most part, of the large night-flying moths and beetles ; on one occasion, however, I found the stomach of a Whip-poor-will filled with the large female wingless ants, which could only have been obtained from the ground, and in all probability in the day time. The common June bug is a favorite article of food with both these birds, and as this is a very destructive insect both in its larval and mature stages, the birds are entitled to our best consideration for the good work they do in lessening its numbers. None of the Swallows, Swifts or Night Hawks ever under any circumstances take any vegetable food, nor have they any habits that are open to objection of any kind, so that our utmost efforts should be put forth to preserve them and encourage them to build about our premises.

I have heard one or two people state that they did not like Swallows about their houses because they brought bed bugs ; how such an idea got into any person's head is difficult to understand, and let me say most emphatically that there is no foundation for the belief whatever. Swallows, like all other living creatures, have their insect parasites, but no parasite affecting the Swallows will ever trouble human beings.

CONCLUSION.

There are other families of birds more or less directly beneficial or injurious to our interest, but space will not permit an extended notice of each. I hope enough has been said to impress upon the mind of every one the great value of the majority of our birds to the agriculturist.

I have seen some estimates of the amount of damage done to the crops by insects in various countries, including our own Province, and although they usually stand at some millions of dollars annually, I believe they are much below the mark. It is difficult to form an estimate of the yearly loss from this cause to ordinary field crops, because the plants are crowded so thickly together that a large proportion may be destroyed in the earlier growing stages without being noticed, and it is only when the matured crop fails to reach the expected quantity that we realize the fact that something has gone wrong, but unfortunately it is then too late to remedy the matter. In our gardens we can more readily see the amount of injury done by insects, and can take measures to reduce it, but in spite of our efforts the loss is still enormous, and consider what it would be if we had not the birds to assist in keeping down the swarm of insect life. The great trouble now is that we have not a sufficient number of birds to keep the balance between vegetable and insect life in our favor.

We all know that the common cut-worm causes great loss every year in spite of the fact that almost all our ground-feeding birds eat great numbers of them and that the birds that feed among the trees and on the wing destroy very many of the moths which produce them, and so we can easily imagine what the result would be to the crops if these creatures were allowed to increase unmolested by their natural enemies; so prolific are they, that I believe the increase of one season would provide a sufficient number to clear off all the crops we cultivate.

A constant war is being carried on between the insect world and the vegetable kingdom. The laws of nature would keep the balance about evenly adjusted. But man requires that it should be inclined in favor of the plants he cultivates for his own use. To obtain this end it is necessary that we should carefully protect and encourage all the forces that will work on our side against our insect enemies, and while they are not the only ones, yet the birds are the most important allies we can have in the struggle. We cannot very well increase their number or efficiency by any artificial means, but we can protect them from such of their natural enemies as occur in our own neighborhood, and we can encourage them to remain and breed about our farms and gardens. If this was done all over the country generally we should find ourselves amply repaid for the small amount of trouble expended, by the protection they would give our plant life against its destructive enemies.

ACT FOR THE PROTECTION OF INSECTIVOROUS AND OTHER BIRDS.

CHAP. 289, R. S. O. 1897.

HER MAJESTY, by and with the advice and consent of the Legislative Assembly of the Province of Ontario, enacts as follows :

1. Nothing in this Act contained shall be held to affect *The Ontario Game Protection Act,* or to apply to any imported cage birds or other domesticated bird or birds generally known as cage birds, or to any bird or birds generally known as poultry.

2.—(1) Except as in section 6 of this Act provided, it shall not be lawful to shoot, destroy, wound, catch, net, snare, poison, drug, or otherwise kill or injure, or to attempt to shoot, destroy, wound, catch, net, snare, poison, drug or otherwise kill or injure any wild native birds other than hawks, crows, blackbirds and English sparrows, and the birds especially mentioned in *The Ontario Game Protection Act.*

(2) Any person may, during the fruit season, for the purpose of protecting his fruit from the attacks of such birds, shoot or destroy, on his own premises, the bird known as the robin without being liable to any penalty under this Act.

3. Except as in section 6 of this Act provided, it shall not be lawful to take, capture, expose for sale or have in possession any bird whatsoever, save the kinds hereinbefore or hereinafter excepted, or to set wholly or in part any net, trap, spring, snare, cage, or other machine or engine, by which any bird whatsoever, save and except hawks, crows, blackbirds, and English sparrows, might be killed and captured ; and any net, trap, spring, snare, cage or other machine or engine, set either wholly or in part for the purpose of either capturing or killing any bird or birds save and except hawks, crows, blackbirds and English sparrows, may be destroyed by any person without such person incurring any liability therefor.

4. Save as in section 6 of this Act provided, it shall not be lawful to take, injure, destroy, or have in possession any nest, young, or egg of any kind whatsoever, except of hawks, crows, blackbirds, and English sparrows.

5. Any person may seize, on view, any bird unlawfully possessed, and carry the same before any justice of the peace, to be by him confiscated, and if alive, to be liberated ; and it shall be the duty of all market clerks and policemen or constables on the spot to seize and confiscate, and if alive, to liberate such birds.

6. The chief game warden for the time being, under *The Ontario Game Protection Act,* may on receiving from any ornithologist, or student of ornithology, or biologist, or student of biology, an application and recommendation according to the forms A and B in the schedule hereto, grant to such an applicant a permit in the form C in said schedule, empowering the holder to collect, and to purchase, or exchange all birds or eggs, otherwise protected by this Act, at any time or season he may require the same for the purposes of study, without the liability to penalties imposed by this Act.

7. The permits granted under the last preceding section shall continue in force until the end of the calender year in which they are issued, and may be renewed at the option of the chief game warden for the time being under *The Ontario Game Protection Act.*

8—(1) The violation of any provision of this Act shall subject the offender to the payment of not less than one dollar and not more than twenty dollars with costs, on summary conviction, on information or complaint before one or more justices of the peace.

(2) The whole of the fine shall be paid to the prosecutor unless the convicting justice has reason to believe that the prosecution is in collusion with and for the purpose of benefitting the accused, in which case the said justice may order the disposal of the fine as in ordinary cases.

(3) In default of payment of the fine and costs, the offender shall be imprisoned in the nearest common gaol for a period of not less than two and not more than twenty days, at the discretion of the justice.

9. No conviction under this Act shall be quashed for any defect in the form thereof, or for any omission or informality in any summons or other proceedings under this Act so long as no substantial injustice results therefrom.

SHARP-SHINNED HAWK
Accipiter velox

GOSHAWK
Accipiter atricapillus

ROUGH-LEGGED HAWK
Archibuteo lagopus sancti-johannis

SPARROW HAWK
Falco sparverius

GREAT HORNED OWL
Bubo virginianus

LONG-EARED OWL
Asio wilsonianus

SCREECH OWL
Megascops asio

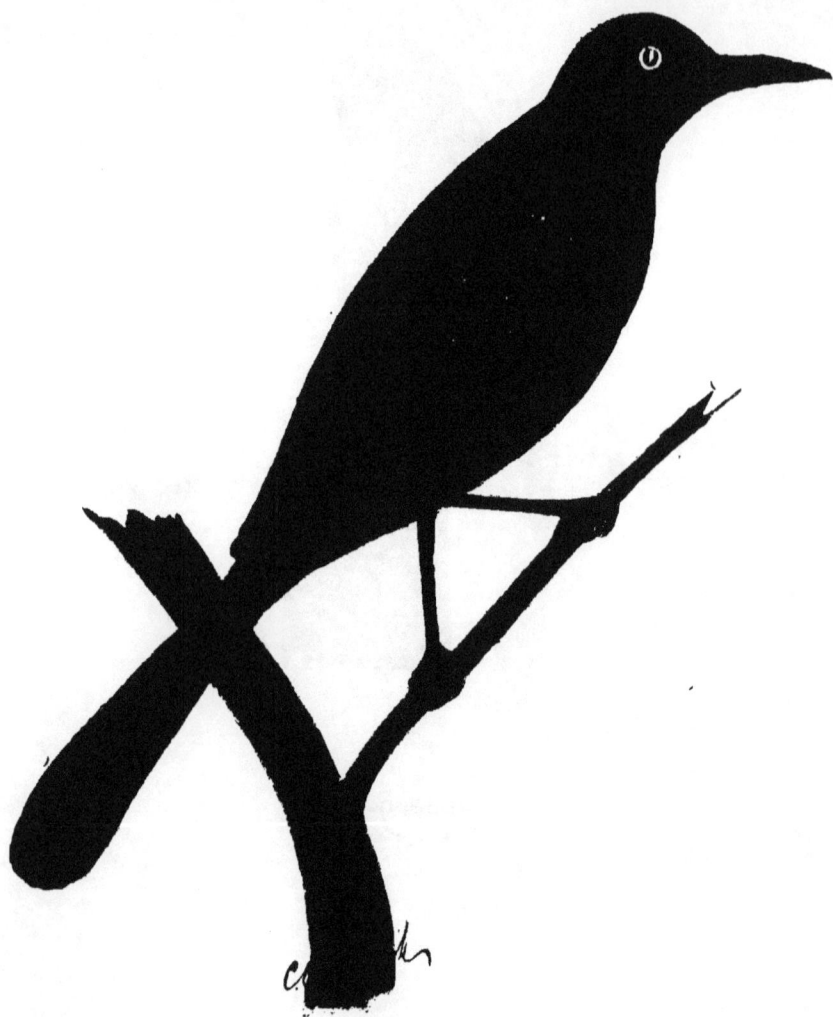

Bronze Grackle
Quiscalus quiscula œneas

COW BIRD
Molothrus ater

MEADOW LARK
Sturnella magna

BALTIMORE ORIOLE
Icterus galbula

RED-HEADED WOODPECKER
Melanerpes erythrocephalus

YELLOW-BELLIED WOODPECKER
Sphyrapicus Varius

DOWNY WOODPECKER
Dryobates pubescens

WHITE-BELLIED NUTHATCH
Sitta carolinensis

WOOD THRUSH
Turdus mustelinus

CAT BIRD
Galeoscoptes carolinensis

HOUSE WREN
Troglodytes aëdon

BLACK-BILLED CUCKOO
Coccyzus erythrophthalmus

YELLOW-RUMPED WARBLER
Dendroica coronata

REDSTART
Setophaga ruticilla

HORNED LARK
Otocoris alpestris

King Bird
Tyrannus Tyrannus

PHŒBE
Sayornis phœbe

WHIP-POOR-WILL
Antrostomus vociferus

FOX SPARROW
Passerella iliaca

SONG SPARROW
Melospiza fasciata

PURPLE FINCH
Carpodacus purpureus

ROSE-BREASTED GROSBEAK
Habia ludoriciana

CEDAR WAXWING
Ampelis cedrorum

BARN SWALLOW
Chelidon erythrogaster